Younger Years

poems

Liz Jones

Clare Songbirds Publishing House Poetry Series
ISBN 978-1-947653-97-9
Clare Songbirds Publishing House
Younger Years © 2021 Liz Jones

Printed in the United States of America
FIRST EDITION

Clare Songbirds Publishing House Mission Statement:
Clare Songbirds Publishing House was established to provide
a print forum for the creation of limited edition, fine art from
poets and writers, both established and emerging. We strive to
reignite and continue a tradition of quality, accessible literary
arts to the national and international community of writers, and
readers. Chapbook manuscripts are carefully chosen for their
ability to propel the expansion of art and ideas in literary form.
We provide an accessible way to promote the art of words in
order to resonate with, and impact, readers not yet familiar with
the siren song of poets and writers. Clare Songbirds Publishing
House espouses a singular cultural development where poetry
creates community and becomes commonplace in public places.

Clare Songbirds
Publishing House

140 Cottage Street
Auburn, New York 13021
www.claresongbirdspub.com

Contents

The author wishes to thank the following publications for originally printing her work.

"Rufus" appeared in the 2012 issue of *The Lamp*, a creative writing journal for graduate and professional students at Queen's University.

"Synaesthesia" published online by *Poetry Nation*.

This work is dedicated to those who have encouraged and supported me along the way.

Bad Faith

This is bad faith,
when I treat you as though you were an image projected on a
screen,
or an objective being without a subjectivity,
or a fact, and you like this.

When I look at you, I should not look at your eyes only,
but at your lips also, at your skin, and your entire face,
reminding you that you are more than a brain to me
and that I like you, you are accepted by me–
you are beautiful.

But you cannot abide by any real intimacy–
always present is the distinction between friendship and love–
and you are indifferent to what this says about you.
So I take you at face value, I accept your word,
and so it continues.

Message in a Bottle

When this is all over, please write me now and then
so that this slippery slope that we call 'time'
might finally have some pause.
Do not be a stranger to me as we are two of a kind
and seraphim are a dying breed.

It was no accident that you were sent here for me and me for you.
It was a gift from on high
to help each through the waste land.

I could have cried when you turned towards me, looking to see,
dark shadows outlining your golden frame.
I, myself, felt nothing inside,
lacking and looking at this blank page so to feel.

Beneath every winter is a warming heart and a heartless spine.
Books are bound but broken inside and unused clutter sully a
bright mind.

When this has all come to pass, read, reflect back, and write.
The minutia of the everyday has come to spoil my vision—
wish me godspeed in its express renewal.

In Your Absence

You came here with a message
and pretty things undone in hours of misery
while I await the sure dance of the undead.

A tide of plenitude promised in silence,
but weary-eyed and accompanied by pure pangs of regret for an
act uncommitted,
I retreated into the fortress of the known.

While all was held up to the light,
there was always the 'should',
the 'ought' of obeisance for the bleary-eyed, stumbling light.

I meant well but couldn't quite express it.
To end and say "no, that is wrong,"
so self-defeating,
but to qualify my meaning, so damned right.

We never quite mean what we say,
and can never quite convey our meaning,
but to feel the undaunted fortress of light–
a symbology of truth.

A window crack awaits your steady presence.
Agreement was always relative to the knower,
but the meaning, he says, was always absolute.

With Your Love

With your love,
I might flourish within,
I might grow to the height of a king,
bathing in the sunshine of that autumnal gift.

Effervescent effusions of emotion would be quelled
in a love that is fully realized.

The flux between yesterday and today would vanish
and the tumultuous waters of existence might calm.
The subject might be stabilized to a new point of routine regularity.

Just as the pulse quickens and then slows after the long lost race,
so, too, would the subject be stabilized by the advent of your love.

The Thin-Tinned Soul

A tide of panic percolates in the heart of a long-forsaken friend–
cracked etchings of a thin-tinned soul, barely half grown.
From barren surfaces arises a silver, multitudinous screen.
Kaleidoscopic and shifting, it tears a piece of me every time it
moves.

And the pendulum swings, the impending promise of my demise,
it moves a little closer to me every time it sways.
Every time I breathe, I fear this slowly encroaching jungle will
invade my senses and smother me to sleep.

Inky night, glowing eyes pierce the silken screen
and a creature arises from a slough of black misery.
Wanton and shifting, it swings its head from right to left, its eyes
fixated upon me, concentrated.
A deep purr issues from its throat and its eyes slide past me to fix
on more desirable things.

Feign would I be to describe all the things that light upon a weary soul
in the eerie moments after dark.
A new sun rises, and the dawn pours an embryonic beauty upon
your wicked face.
Fye on my soul, misery passed in hours awaiting the rise of the
dilapidated daffodils from the frozen earth.

With first spring stirrings, their slow ascent shames weary winter
into submission.
In Hades, there are many things of which we cannot conceive.
In heaven, held on high, there exists a faint, gilded plane,
an etching of a tile from your coat of armour.

In clamorous terms, we fill our folly with rotten apples and arise
insensate from a foul sleep,
dumbstruck by the brilliant speech of tinkering dolls and hand-
some Hansels in retreat.

Across linear planes, I move in my seat, so that you might have a
new angle on me.
This thought has shifted to the nth degree and it is off
by 14 degrees.
The round, gesticular motions of a shape whose voice has been
muted fills my peripheral vision whilst
my gaze falls upon my feet.

Octagonal View

In a continuous state of flight while never landing
lies real living.
Erasures of uncertainty, fingers search silently for meaning.
Cold unease encompasses a recumbent form
riddled with old age and excess–
signifier of the ultimate acceptance.

The sudden, slow, downwards spiral–
tailspin into the obscure.
You made an appearance,
your light was exposed.
Now vultures have come to prey,
fully cognizant of their effect on you.

You couldn't have been quieter,
but they lay in wait for you.
Heard your step in the dry grass
and crept into the abscesses of thought where
they aimed to thwart your determination.
Now they frolic, missing in action.

Walking into the light, sunbathed, you
forgot the world peering in on you–
or had the sunlight blinded those effervescent rays?

In the light, they still are known.
The prophetic future, Tiresias' blindness, awaits.
Notwithstanding, he who knows least sees too late.

There where the moonlight bends its rays,
you fade into the octagonal view.

Sad Dolly

Locked away in catacombs of memory
crowds throng on all sides
My god
relentless struggle of the day world
the place of calm amongst the rocks
amidst the flux of the world
in your own mind.

Retreat from places with no guarantees–
affixed solutions
to internal dissolution.
Deprivation from people in private places–
the poverty of spirit–
I offer no social or economic utility–
alienation.

Time, the wayward force,
steers in directions leading off course.
Thrown back on oneself, nothing to lose,
my god, please help me mediate my reality.

Florid Manuscripts

Fear of interpolation by a foreign source,
alien, suspect, body, concubine, suspicious rogue.
Cocooning in solitude,
solace from the external world,
too complex to be tarried with–
too omnipresent to lack meaning.

Dig into the depths of the unknown.
There you will find the latticework
as foreign atoms ionize in a celebratory dance
and a chrysalis is discarded.
Walking along a predetermined path,
find the love that was lost at last.

Paradise

No paradise here, or over here, but look yonder!
Green meadows and golden pastures,
Ox-blood red,
      burnt orange,
        amber,
        f
      a
       l
        l
         e
          n
         leaves.
In piles, lying in clumps, beside mounds of hay,
between rolling hills, which stretch for miles on either side–
crests and troughs from here to there.
Islets immerse the fractured gaze that twins into two eyes
before it falls short and sees no more.

But the sea in between! Only the primeval imagination could
capture the 'discovery' of a 'new' world,
Crusoe's countenance in New Guinea and Deirdre's separation
from an abusive mother.
Pericles was thrown this way and that before the Gods washed
him ashore, finding 'freedom' in bondage to Simonedes, born
to be bound to both Thaisa and the kingdom in exchange for
his life?

Vengeance knows its price, twice thrust in spite against those
who will not bend to their will.
But neither kings nor beggars are we, so cringe not under thy
Lord's will.
The supplicant never received relief from mighty man, nor from
the peacefulness of passivity, but from a sheer act of grace.

Hope

Moving beyond the chasm, the great unknown,
logic forgone, the prodigious leap of thought hastening to
conclude, excluding the middle,
perplexed by the unsaid, you quake.

Let us then comb through the threads of chance, remarking
upon the events that have passed,
bolstering confidence with reason, speaking and statements
strengthening resolve,
dimming the hush of the mundane.

Reviewing the first course, I started in my Sunday clothes.
Upon the second, incense seemed to waft from multi-coloured
fabrics.

In repose, a peacock was accepted at the dinner party. Now we
await the prophecy of a pocket watch,
mysterious harbinger of things to come.
But let us now marvel at the spectacle that we entertain,
embellishing colours with names.

When the night ends with ethereal down, we shelter the horses
and extinguish the lights.
Perhaps when eyes open with deliquescent youth, we shall
delight upon the path we choose.

Synaesthesia

Tender night of the heart-warmed soul,
Around buxom hearths we sat, warming.
What was to come unknown,
Shadows flitting over shaded eyes,
A black and light kaleidoscope.

Twice I felt and thought the same, taming the extreme.
Saddened, abiding, matured,
The flame was captured, extinguished.

Then the medieval choirs rose by soft degree,
the mahogany pews crumbled, decayed to dust,
the beginning, inchoate,
shaping formed masses from cinderblocks.

Shades blurred, shifting into sound,
pitch then high, then low.
Thought, word and deed–a manifold penumbra of impressions–
each associated with a particle,
a thinly gripped, finely felt junction, unplumbed.
Heralds hearkened: the sweet bloom of spring.

In God's Hand

Spinning dancers, not far from Derbyshire,
squirming, wormy bundle of love.
When the mind clouds over with earthly care, the face pales,
images blur into one another, indistinct.

In Earthbound leather, you circle x's and o's–there are no patterns.
Concentric spheres shift in relation to this perplexed planet, sitting
off-kilter,
the axis a fulcrum of this world,
a shifting, fixed point.

The four winds alter their course.
Orion oversees Atlas straining under a weight he will surely en-
dure.

Turn to our Indigenous roots. A tortoise supports this world–
a contributing, necessary participant in this embossed volume.

Down the road, orchards sit, to be plucked by the ready.
Here is an ancient spirit, neither warrior nor hare, thought shroud-
ing action–
but what is the life unconsidered?

"Delay deadens all," quoth he, grizzle-bearded, steady voice,
hands all knobby, on knotted cane.
Time awaits not the unsure footed and so, slouching forth,
we make our amends, cut our losses and breathe decision into the
lifeless vessel.

When the vapour of thought dissipates in the grim wash of day
and Concern is pushed from centre stage, the shadows lengthen
and night falls.

In the eye of the flame exists a fleeting hint of all that could be,
fading as quickly as the players make their appearance and depart.

Held Captive

As the wind shuts the door
and snuffs out the light,
the mind moves silently
in the dark dance of night.

The heart strings plucked
in a soundless hymn
playing one's ownmost instrument
through the labyrinth within.

Through a window,
the whole world shifts,
blurring hue and shade,
tone and pitch.
In my mirror reflection,
I see my better self,
remembering who I am,
departing every day.

On a beach, discordant voices
mute the symphonic surf.
Unable to concur,
we tear reality apart.
Here driftwood washes ashore,
a remnant of the ship where you said, "I would."

Promises better kept in another time,
fall apart before human beings are tried.
The young pass old, the old young.
Under a trellis with you, I laughed,
meaning I won't.

When the days are shorter, the nights are long.
With all in bed early, I rehearse my song.

A Return to the Senses

Shall you re-assume your rigid, arcane structures,
Timelines, deadlines, hierarchies of knowledge and incorrigible
truths
deaf to the blows of Fortune in the everyday?

In the city bus, passengers assemble, tum-
ble, sit and fall,
forks in the road suggest indecisions,
revisions checked by the osten-
tations of Time,
rotating on a flattened plane.
The epistemological quest for certainty in the
ordinary thwarted,
buffeted with every blow.

So, tunnel-visioned and bleary-
eyed,
she accepted that there could be
other kinds of truths–
was absorbed by her surround-
ings.

Bird in a Cage

Pausing, here, before walking through the door
That you hold for me
Aware that if I do not accept this gesture
The frame will be barred to me.
And so, I lift my sorry feet–
Sorry, indeed, that such a thought has occurred to me.

Each is constituted by the other in this world.
Being for another sex, being to be?
Such grief in existing for the other, but such grief not to be.
To be let alone, fallacious thought, indeed–
allowing oneself to be left alone permits the vultures to feed.

This is not a game and they are not here to play.
You men quit rooting for the 'winning' side!
I don't play chess with birds of prey–they are not rule bound.
When Thor raises her thunderous voice,
Their nets and snares will come crashing down,
Hopelessly entangling them in their own designs.

ABA ABA ABA...

Your seminar exploded like a villanelle in my head. All I could hear was the silent refrain "you must abandon your wish for mastery." It having disappeared long ago like a cup whose crack cannot be identified; I still do not understand the reason behind your obsessive rhyme. And, in sacrificing this desire, I was made worse for the wear, stunned by the ever-shifting landscape of change. The message now imparts little meaning to me...little meaning to the point of absurdity. And the chipped cup could not be repaired.

The Drought

Dreams of a restless night
Turn to thoughts of a living day.
Cold hearts and clutching hands
Carry stark images of a fixed and vexatious past.

The content places and plans of yesterday are undone.
The plans for the future: once a bonafide venture, an unrealized
possibility, now controvertible.

Starched linens and good tidings–
A piece of furniture unused.

Green cedars–
Shades of green–
A swing chair on a veranda, weathered but unused.
Husks of dried corn stalks
And dry autumn air.

In Reading

Lunar eclipse–nothing,
To deny what is real
As an apprentice at the wheel.

To be married in a year
Was her greatest fear,
But vulnerable, indeed,
She chose the separate sphere.

And now embroiled in language that she does not understand,
Contemptible.
Washing surfaces and rolling up sleeves,
Down on hands and knees,
Is this the way to proceed?
Heaven's hell is at the gate
and the dogs are barking.

The Armchair Historian

"Dreary me, fetch me a tea.
Hurry, hurry, the leaves are on us.
I have left my dear in the city of Leeds,
and am saddened by this mass disaster.

I am saddened soon, but am saddened no faster,
than when I witnessed the Reign of Terror on this fresh slick of
plaster,
and in addressing sad fact, I am not fit for visitors,
but perhaps I might see you this afternoon."

Time slips by all too soon,
For the Vacillating Spinster with a clock ticking in her room.
Tick, tock, it all slides by
And in sliding slithers to a darker time.

"A darker time, really," you ask "than in times like these?
See how Reason wipes out Superstition with its sleeve.
Rule by a dictatorship of inductive probabilities,
derived to suit the times in which we live,
is this so preferable to a world in which God succeeds?"

"Ah," I answer, "it is not ideal,
but each must be a person of their time,
so, you best shut the fables of the figurative mind."

Metronome

Space, time, the conditions of experience,
the metronome hot clock sits ticking above the door
a time bomb waiting to implode
keeping time even when we aren't.

Staccato wooden soldier
stiffly moves across the acrid floor.
Each jerk a precious second slips by.

9 horrific minutes for a horrific act,
In less than a minute someone's life is forever altered.
Mid-air collisions, saboteurs, criminals.
Stealing someone's time, robbing them of their lives.

Who said human beings could play the devil?
God, why do you let them?
Ruining the natural,
Sick, twisted curbed machine,
Sitting crucified, ticking in the corner of my room.

A Cambridge Relation

Out my window, I see a figure beck and call. When I am
scared, it is ominous.
When I know I have nothing to fear, it is friendly. When I
wonder just what it is, it is inquisitive.
When I wish to meet it half-way, it is willing.
And when I find that it was just a door post, it remains just that
—a door post.
Nothing has been divined and nothing is learned.
And when the subject rebounds from their object, they will find
that it is as it always was—a door post.
       And this relation is absolute, and it will not change.
       And yet we want for more than posts.

The Seer

Sifting through the posted new articles
intuiting God's designs synchronistically,
from the sensory to non-sensory,
the diviner embarks upon the task of turning water into wine,
seeing His message in the mundane, thatched houses,
or in the formidable event that strips away the thin veneer of
sociality, leaving the human being exposed to the elements,
without a lotus leaf for coverage.
Whether in finding God's work in nature,
or seeing His supercilious
acts happenstance, Wordsworth's poem upon a strand,
the flight formation of birds, military V, upon a cold autumn
day, it is for you,
one of the fortunate few who now dwells upon this night ship,
feeling the steady lull wrought by the ship's hull as it cuts
through the currents.
Moving by way of indirection, scuttling sideways like the crab,
towards the human's final destination: union with the heart's
proclamation

The Journey

Upon an open sea, this little lifeboat sits,
like the lotus leaf, belying its underpinnings.
Stability, permanence, a fleeting thought for its survivors,
huddled for warmth,
distilling water through a sieve to feed the famished.
Rations: old tin cans, cracked open with a hatchet and the
occasional fish,
makeshift fishing rod.
Water frolic forbidden by the elders lest the children awake
what lies beneath,
efforts to survive whilst nature encroaches on all sides.
At night, lantern light illuminates this little houseboat, hoping
for a search party,
some respite from sheer immediacy.
Yet, the mind, working overtime, thrives on unanticipated
resources.
The elders survey their progeny, like so many tendrils cast
beneath it.
Stone faced, always on the edge:
reaching the final destination determined solely by the journey.

The Wind

After a soul connection, one exists on the other side,
as the dove subsists, soaring on the wind,
beating not its wings.

The heart suspended in time grasps not for little things
is seldom heard until the bird enfolds its wings.
Meter for meter, air for air,
At times not swift but always fair.

The Initiate

In "the Bacchanalian revel...no member is not drunk."
When not drunk on wine, one is not a member of the club.

For in these revels, Bacchus attends dreamers and whispers
sweet things
that make them believe their waking life was but a dream.

While visions false but true seeming,
Bacchus is nonetheless exactly what he appears to be:
a god of wine bent on changing a dreamer's reality.

Firefly

Radiant espionage
Night entrails, flickering gossamer wings,
Frolicking beneath lamp lights, little helicopters,
Spin jets, jumping and dancing, jubilant.

Youth, shrieking, jumping, grasping for
Magical phenomena, ephemeral clouds.
Energy descending the trio,
Radiating from the eldest,
Immersed in the moment,
Cognizant, the Second,
Of the unknown.
Fear of the unknown.

But dancing fireflies in lamplight,
Nest of nature and home,
Rapunzel's wall,
Bright light.

At the Speed of Light

The light hit the cobblestone
and then was a million miles away,
off to another world from whence it came,
onto an island universe that I will never see.
Wheel of fortune, wheel of light,
How quickly your spokes go around,
a common sphere, a common cause,
now here and then you are gone.

The Domicile

Who heeds the lark's arising?
The owl hears its melodious strain,
when fuchsia, intermingling with ox-blood red,
sees red give rise to white.

Figures diminish with mired age
as the child's surely grows
and flowers fade with severed stems
as the river overflows.

Change discrete, within, cuts the world in two,
life fades with weary age and the mist with breathy dew.
On hanging boughs, we await the day that life might
incorporate two.

Outside

Assembling a piece, I dwelt on one single task and time
stretched endlessly on.
In time that is no longer my own the precious moments slip
away privily.
The day has external value in the eyes of others and this, per-
haps, forges a common path?
The road least taken had become overgrown by grass and this-
tles and lifeless dandelions nestled in amongst those fresh and
alive.
A rusted blade for felling birches, a cutting block, and old wires
entangled amidst nature's beauty.
The rise and fall of the wind and the rustling of leaves, waver-
ing and indecisive, infused the landscape with strange vitality.
All around me was dancing light illuminating natural form.
And then I was recalled to my task.
And the cat was just a cat again, a pet outside a stranger's
home, and the tree was just a tree, and the bird watched me as
though I were strange.

Red Roses

Red roses, black vases, still life cut-outs, thick air.
Motion paused in space, the melodramatic dip of she as his
mouth grazes her throat.
Ephemeral lull in the darkened day, the Gypsy waltz seems to
fade, drifting from
corridors everywhere, onto a darkened stage.

O symphony of the machine whose wheels fail to turn,
O earth blown bliss, the sun rises high to illuminate this world.

An Ode to Friendship

I want to lie in bed with you and watch a movie, dimly.
The din pitch of the machine desensitizing my frazzled brain.
I want to see shapes and colors float across my ceiling and walls
and vaguely know that they are projections.
I want to hear voices turned down, non-distinct and non-
threatening,
unrelated to me and my meager existence.
But mostly, I want to lie in bed with you and know that you are there,
but that you don't care and would go on with or without me.

An A Priori on a Winter Morning

When I opened my eyes, it was there before me
and there was nothing that I could do,
      nothing that I could see.
White snow had drifted in my window
and the outside world had pulled me from my sleep.
Lost in the fragility of a youthful mind
that sees no separation between "they" and "I",
this child had forgotten the passage of time
and the majesty that awaits the unassuming eye.

Friend or Foe

She appears at my door, dethroned, not to be trusted.
Fearing, wanton, up in arms, now that I am disrupted,
the reasonable wishes that it was not so for the Hobbesian human,
holding fast to their property.
"What's mine is not thine," says I, but in the wilderness such
distinctions are not made.
Play your cards close to the vest without bringing every motion
to a rest.
Like the ice-skater, pulling limbs in close, picking up speed in a
momentous whirl,
until it is safe to extend.
Prima facie, the evanescent glimmer of a fixed performance.
For those who follow the dictates of reason, adhere to this distinction between sense and the senseless.
"Leave it," the dog owner says to the dog fast approaching a
stranger.
We see the humor in such utterances where
consideration of the audience's response dictates what is said.
Heard out of context, the human being is not distinct from a
piece of dirt.

Rufus

Rufus on the roof again,
roaring like a ruffian.
Roundabout the rim rips this wry wrangler,
ripping up rickety split tiles from the roof top.
Rallying his remaining forces, he makes a final round,
and wakes the rancorous neighbors with his racket.

Sweeties

Rounded tummy and limbs, pink-ribboned hair,
the empty basket on her would-be motorcycle longs to be filled
before embarking on a day-long journey with Ma,
good-natured, wry, ready for ranging requests.
Streamers await the wind of a hurried journey to the corner
store.

Summoning her strength, muscles taut, heart pounding,
her mouth forms a sibilant, first sound of the forbidden fruit,
then a glide and monophthong,
followed by a stop, two vowels and the final fricative.

Forever venturing upon the unknown, tremulous, trying lessons
of courage.

The Porter

I thought I caught a glimpse of the porter,
walking down the street,
but his hat was over his eyes and I could not see beneath.

I thought I caught the porter's green, green eyes as I queued for
the show,
but I couldn't be sure if it was me that he saw, or the guest
behind–I still don't know.

The porter ushered me into the lift, but it didn't take me to
where I wanted to go.
I chanced to see him exit the boiler room next door–where he'd
banged around for 15 minutes or so.

I don't see the porter much these days–he doesn't come around
here anymore.
But then again, I don't go where he once was either, and it's
hard to say when you just don't know.

Breathe

Tuberosity, daffodils blooming into valence
Thick stalk unveiled from rich earth.
Bulb, roots, the messy underpinnings–
Genetic replication in every strand.

When seeds blow in the wind
Marsupial swamplands quiver in anticipation.
Birth in the breeze cooling fetid skin,
Plantain amulets, thick corn stalks cleft from cobs,
Succulent pomegranate seeds, fruit of desire.
Perspiration of satiety, cornucopia of desire.

Immanence

Sun blaze, the rat race of the day to day.
Black night, blue black embers enshrine the dreary head,
soft hands embalm the bleary eyes.
To see outside the little room,
inspiration ripples within the gloom,
to see outside the fetid day, with heavy heart and faint restraint,
the life within must surfeit one thousand ills worth preserving.
Until the nought subsumes the ought, day will reign, night forsaken.

Moonchild

As a shaft of light seen through a prism,
insight emerges, proffering incorrigible truths,
quantifying intuition's understanding of reality,
which, while readily felt, was hitherto unexplained.

Circumstance, when reduced to a fine line, is made plain,
In so doing, consider what we take away–
the mist and vapours of the unthinking brain,
feeding on possibilities, whittling the time away.

Love not left behind, but truth conveyed:
"For now we see in a mirror dimly, but then face to face."

Scrawl

Alien unto myself,
tree limbs sway–
sound resonates from a hallow place.

The interface, a chip
in the holistic self not there.

Bombardment of sense impressions and thoughts–
maternal body, external corpora,
not existing beyond singular mind.

Referents refer back to the speaker–
the writing is on the wall.

The Imago

Trapped in memories of ardor, recollections of flippant flirtations.
We had refrained from addressing the concrete to play out fantasy.
After the clock struck 12, insignia between two realms,
the transformation of a six into a seven year,   ˈ
the social impulse became rationalistic, each embodying the role
of both scientist and mystic.
The seraphim returned, a golden light shining behind spectacles.
The return after a fevered kiss: the Tao of Steve was not achieved
through retreat and return.
Cinderella lost her glass slipper, but it won't be returned.
One moves on by assuming a new object.

The Slippage of Identity

If I could write a separate life, pseudonym, anon, anonymously,
I'd pen my life's journeys–
Yet, barred, refracted, restricted, folding into
one-dimensionality, feathers become fetters.

Life's short letters, soft-bound leather–
journals alter into phonebooks; phonebooks fail to form directories–
ruminating upon the past and present while concrete reality
slips away.
Structures dissolve into chains of data, the whole is lost to the parts–
with the rational mind, the world is torn apart.

Yet, in having control over affairs, aware, instep, insight, all is
not unwell in the world.
Becoming more aware, activating the visual-spatial spectre to
see what others see–
all is forsaken when studying to be.

How to unlearn the tutelage of learned thought,
Unsubscribe to a dream, assign prescriptives to steadfast
minds?   12

He loved too late, too hasty, rash and intemperate in youth,
Werther did as he pleased.
But fear not the riff-raffs,
Lifeboats sail away to Elysium islands where feeling is not
shunned.

Unbind me now to resonate with today's tune.

The Stranger

Stranger, stranger, in the room,
Thy shape my eye cannot subsume.
Waiting, watching, the stillborn air,
While thought inert holds thought within.

Within the wee hours of the night,
spectres walk this earth with no one in sight.
Vagrant thoughts come to mind,
so that I may repose and pass the time.

To See

You want to sit on a highchair surveying the parade, aloft and
separate from the world.
Isolated towers tell tales–the guards viewing the panopticon,
seeing or seen?
Herein lies the power to let others view: accept the power of
viewing.

Under the facade of court jester,
the human being controls an image whilst relating a presenta-
tion.
Enigmatic figure: crying clown or joking trickster?

Seating plans at private functions and in public institutions,
necessary arrangements of physical space–
together and apart: complimentary pairings and separations.
Therein lies your power over the haphazard.

Holding a cocoon tenderly in clammy palms,
waiting for the caterpillar to transform into a butterfly.

Drunken Yahoo

Drunken yahoo crawling amidst the sprawl of garden furniture,
hooting and hollering at the stray cats and passersby,
disrupting the peace of the terrestrial plane
and the dignity of the domicile.

Swath of drunken misdemeanours, petty household saboteur.
The four walls of the interior cleaved out during a chainsaw
renovation
exposing bare beams, an unforgiving neon light.

When locked doors deter entering rooms, a pickaxe will do.
Incommodious to its inhabitants, most abandoned ship–
a few confine themselves to the comfort of a rocking chair.

Numb

Frayed wire on a line–
noose feed a spread–lay out your discontent–
hold yourself on the line,
hostage to no one but yourself.

The underbrother feeds limericks to the children
to spell out their inner selves,
lying unattended in the wake.

Honey drips off the overman,
the social self, unsustainable, a fractured lens
that crumbles when questioned, collapses when pressed.

Held up by all fours, is it man or beast?
We rise by day,
by death desist.
'Tis better to know this than to feel nothing.

The Beleaguered Player's Appeal

"Empyreal emperor, the troupe could not subdue
clamorous echoes arising from the crowd
dissolving our performance for you.

Heavenly king, half stayed, the others flew,
the goodly not conceding to an act disavowed.
Empyreal emperor, the players could not subdue.

Left bereft, in clouds of dust, the vivacious lose their hue.
Behind iron bars, they brood the past, downcast eyes, heads
bowed,
dissolving our performance for you.

With vacillating spirit, he who absconded and eschewed,
turns traitorous wit and cunning eye against the pledge he
vowed
fracturing the band, wresting them in feud.

Our stage has collapsed and cannot be renewed
unless the petitioners' voices be permitted to speak aloud,
dissolving our performance for you.

Saddened though I am, kneeling on this pew,
entreating you most graciously, please be not proud.
Empyreal emperor, the troupe could not subdue,
dissolving out performance for you."

Life Isn't Like a Mathematical Truth

Our meeting here was just a mistake–
I caught a glimpse of you as you walked past the room.
I planned on more than I could undertake
And so I was left with more than I could consume.

I was never told that life isn't like a mathematical truth.
The 'rules' are being bent as of late.
Fortuitous plans made in the prime of youth,
Fortuitous plans without fortuitous fate.

The lowly ant crawls about and builds its castle amongst other ants.
It is burdened by more than its being would seem to bear.
It does not question reality so long as it recants–
Its nihilism will not lead to despair.
Holding close, walking a fine line,
its task is commensurate to turning water into wine.

Filaments of Fire

Wind at my heels–
Black plume of night.
Lasso behind the wheel
A shot out of sight.

A picture is formed from fragments of life–
A deer is notified of never-ending sight.
All in all things considered
bear happiness and strife–
and no in nothing can compare to this life.

Dragging this Carcass from Day to Day

Here be I, dragging this carcass.
I move it–it does not move me.
Here I be, crammed in a seat.
Happy to be self-moved, a self-starter.

There I am, a morbid sense of humor,
begotten after a weekend of jollity.
Now I feel no frivolity,
But only estranged light, and stark truth.

No More Speculation

We need signs that speak to ordinary human beings.
No conditionals, ifs, ands, or buts.

"You open product, you buy product" reads the sign in the
shopkeeper's store.
No please and thank you on this sign–
pleasantries being dispensed with here for lack of an
appreciative audience.

And seriously, no conditionals–
they were never properly understood by ordinary man.
And besides "signs with too many words make you sound
stupid to them," said the shopkeeper to me.

I and We

I am not I; I am we.
See how I substitute thee with me.
We are not many, but all is one,
and I am I and not I until my days are done.

The "I" that is "we" has disappeared,
And all that is standing is absurd.
What's the same for one is not the same for all.
We were better off before TV and mirrors.

Silent Conversations

Film noir, camera obscura,
Byzantine sculptures, the partial undress
Of a Duchampian "Nude Descending a Staircase."
L' esprit de l' escalier, the cool afterthoughts.
All said and done,
converse freely, no regrets.
The free exchange of information–
fine importer of ideas.
Fabricated with golden silks of the ancient world,
incarnate, sensual, a sensuous being
given in concrete immediacy.
Non-neurotic, no glitch in the machine.
Only attention to detail is appreciated.

Airing Your Dirty Laundry to the World

I don't know you and you don't me,
but your public announcements told me everything.
I've seen others in this situation–I feel I know your past.

It's gripping, it's touching,
it's sad–it didn't last.
How could this be, a star in the sky?

But a star in its place is nothing in another's sky.

The outward move after a failed pass,
an attempt to distract, some attempt
to gain back.

All this is fine, but has become apparent to me
simply by being one of the few in your proximity.

The Thing You Don't Want

Don't make an appearance
Without any cause
The cinders that will be left
Should be reason enough for pause.

And although I might reconsider
The truth of your claims
The painful fact of this should be reason enough for pause.

Parallel Universes

You are outside the door,
But I thought you were overhead–
In my situation–
In the other washroom. The thought
Meant so much to me–
The communication held intrinsic value,
As reflections of one another's situations.
It spoke volumes to me, something about a
Simultaneity of presences–
In my stoned, sorry state.
In one place, at one time,
Along a vertical axis.
The door opens
And I realize my mistake.
Hallucinations brought about by imbibing
foreign substances.
The dynamics of projection
and recognition
Now clearly seen
To have been at play
Both in my confused state
And in my moment of discovery.
And now I see how much this really
Means to me...
Not as mere reflections of each other,
But this...this being here with you now:
This being an
opportunity so rarely afforded–
I can really get excited about it!
I could hear your voice in the other room.

The Vulture and the Man of the Salon

Get there early or you might not get there at all,
If you rely on the normal course of nature, you are doomed to
fall.
Pay heed to instruction, the voice in your head,
Blind recourse to action will end with you dead.

Now, recall that the frame of time is a life,
this you were given by God and Jesus Christ.
Distractions insignificant, a waste of your time:
a life that depends on them has no reason or rhyme.

To be mutable or fixed is a wayward force–
for the former lacks principle, and the later lacks choice.
Hither thither with the wind flies the would-be man of the salon,
A life bred on gossip, sealed by the vulture's talon.

The vulture, desperate to compete with what it would destroy,
has failed to learn that negation of a friend leaves them with
nothing to compete with.

Author's Notes

"In Your Absence"

Line 9-In his introduction to *Phenomenology of Spirit,* J.N. Findlay refers to Hegel's portrayal of German culture in Hegel's chapter on culture in this work as reflective of Hegel's perception that the German nation possessed a certain kind of sturdiness, but also a lack of polish and refinement when seen in comparison with certain other European cultures. This comparison is based on stereotypes that are certainly not true, especially in the globalized world of today. Nevertheless, one could certainly apply some of the qualities that he mentions to oneself, seen in the absence of any kind of national affiliation.

Line 11-T.S. Eliot eponymizes the moment of naming the unnameable in "The Love Song of J. Alfred Prufrock" when he imagines relating some grandiose scenes to an immovable audience who rebukes his self-expression, his performance, with the simplistic phrase "'That is not what I meant at all. / That is not it, at all'" (Eliot: 16). Later, a chiasmic repetition repeats the refrain: "'That is not it at all, / That is not what I meant, at all'" (Eliot: 16). In general, in the first part of this poem, the speaker, like sense-certainty in Hegel's *Phenomenology of Spirit,* self-consciously struggles with saying what he means, on display and exhibited for the reader.

"With Your Love"

Line 10- A similar metaphor was employed by one of the metaphysical poets.

"The Thin-Tinned Soul"

Line 12- This line is meant to sound *Hamlet*-esque.

Line 13- In his etymological investigation of the German word 'unheimlich' (unhomely) in "The Uncanny" (1919), Freud quotes K. E. Georges who uses the phrase "'in the eerie night hours'" (Freud: 125).

Line 17- "There are more things in heaven and earth, Horatio, / Than are dreamt of in your philosophy" (Hamlet: 1.5. 167-8) says Hamlet. Freud cleverly re-arranges this line in his work *On the Interpretation of Dreams* (1899) as editor Hugh Haughton notes in his introduction to a work entitled *The Uncanny* (2003), a collection of essays by Freud, named after his signature essay, "The Uncanny" (1919). Haughton quotes Freud as stating "[c]reative writers" "are valuable allies and their evidence is to be prized highly, for they are apt to know a whole host of things between heaven and earth of which our philosophy has not yet let us dream" (Haughton: viii).

"Hope"

Line 2- Hegel discusses a concept similar to the law of excluded middle in his "Preface" to *Phenomenology of Spirit*, in section 27, where he polemically states that "[t]his process of coming-to-be" "will not be what is commonly understood by an initiation of the unscientific consciousness into Science" and certainly will not "be like the rapturous enthusiasm which, like a shot from a pistol, begins straight away with absolute knowledge, and makes short work of other standpoints by declaring that it takes no notice of them" (Hegel, "Preface", *Phenomenology of Spirit* (p.15-16, ¶27)). He articulates variations of this same essential idea elsewhere in his "Preface."

"Synaesthesia"

Line 10- A similar image of pews melting is employed by one of the metaphysical poets.

Line 14- The monarch, a figure of absolutes in Hegel's *Phenomenology of Spirit*, has a quavering voice that erupts in a similar manner.

"Bird in a Cage"

Line 16- This happens in one of Chaucer's poems from *The Canterbury Tales*. And, of course, "nets and snares" is a common phrase that has since been cut adrift from its origin in the Bible.

"The Armchair Historian"

Line 2- A similar line (a reference to "leaves" approaching) is mentioned in Sir Philip Sidney's "Astrophel and Stella".

Line 20- A similar drama plays out between Reason and Superstition in Hegel's *Phenomenology of Spirit*.

"The Seer"

Line 10- Imagery from Steven Knapp and Walter Benn Michaels' essay "Against Theory" (1982).

Line 12- "[W]e happy few" is stated by King Henry in his St. Crispin's Day Speech, from Shakespeare's *Henry V* (1599).

"The Initiate"

Line 1- This is a line in paragraph 47 from Hegel's Preface to the *Phenomenology of Spirit*.

"The Domicile"

Line 4- This poem employs a similar meter and rhyme scheme to Christina Rossetti's "Who Has Seen the Wind?"

Line 12- In Lana Del Rey's "Video Games", she says "They say that the world was built for two".

"Moonchild"

Line 10-1 Corinthians 13:12–also cited by Thomas Aquinas in On Faith and Reason.

"The Slippage of Identity"

Line 12- A moment such as is found in "A Dream Deferred" (1951).

"The Stranger"

Line 1- The repetition of "stranger" on the first line and some lines of trochaic tetrameter resemble William Blake's "The Tyger" (1794).

Elizabeth Jones is a Canadian poet. She grew up in eastern Canada. An avid poet and prose writer, she also helped judge a literary contest in the winter of 2014. Jones enjoys living in Canada and hopes to continue working in the literary arts. Her work has been published in *The Lamp* and appears online. The following 57 poems are a sample from her collection.